MOOSE

by Elizabeth Andrews

Cody Koala
An Imprint of Pop!
popbooksonline.com

abdobooks.com

Published by Pop!, a division of ABDO, PO Box 398166, Minneapolis, Minnesota 55439. Copyright ©2023 by Abdo Consulting Group, Inc. International copyrights reserved in all countries. No part of this book may be reproduced in any form without written permission from the publisher. Cody Koala™ is a trademark and logo of Pop!.

Printed in the United States of America, North Mankato, Minnesota

052022
092022

THIS BOOK CONTAINS RECYCLED MATERIALS

Cover Photo: Shutterstock Images
Interior Photos: Shutterstock Images

Editor: Grace Hansen
Series Designer: Laura Graphenteen

Library of Congress Control Number: 2021951832
Publisher's Cataloging-in-Publication Data
Names: Andrews, Elizabeth, author.
Title: Moose / by Elizabeth Andrews
Description: Minneapolis, Minnesota : Pop, 2023 | Series: Twilight Animals | Includes online resources and index
Identifiers: ISBN 9781098242084 (lib. bdg.) | ISBN 9781098242787 (ebook)
Subjects: LCSH: Moose--Juvenile literature. | Moose--Behavior--Juvenile literature. | Twilight--Juvenile literature. | Nocturnal animals--Juvenile literature. | Nocturnal animals--Behavior--Juvenile literature.
Classification: DDC 591.518--dc23

Hello! My name is

Cody Koala

Pop open this book and you'll find QR codes like this one, loaded with information, so you can learn even more!

Scan this code* and others like it while you read,

or visit the website below to make this book pop.

popbooksonline.com/moose

*Scanning QR codes requires a web-enabled smart device with a QR code reader app and a camera.

Table of Contents

Chapter 1

Forest Giants

The moose is a giant, **majestic**, **mammal** of the woods. It walks around the thick forests of the Northern United States, Canada, and Alaska. Moose must live near bodies of water.

Watch a video here!

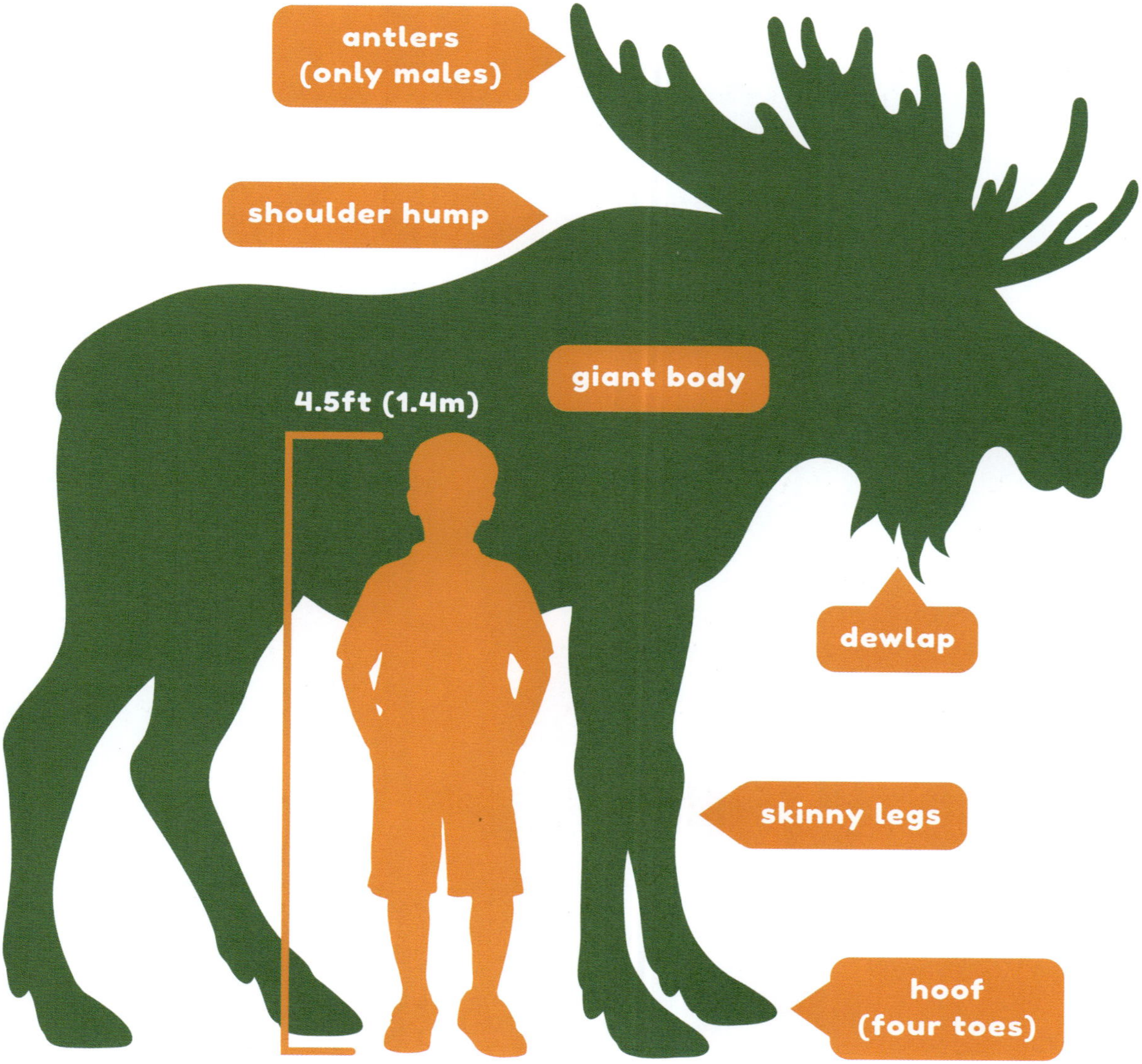

antlers
(only males)
shoulder hump
giant body
4.5ft (1.4m)
dewlap
skinny legs
hoof
(four toes)

Moose are the largest
member of the deer family.
They can weigh up to
1,600 lbs (726kg). They stand
about six-feet tall (1.8m) from
hoof to shoulder. Moose
have long rounded snouts.
A **dewlap** hangs below
their chins.

Moose need to live in cold places because they are big and covered with lots of hair. It is bad if they get too hot. Moose hair is **hollow**. This traps their body heat and keeps them warm during freezing winters.

Chapter 2

The Moose's Crown

Male moose are called bulls. They grow antlers every spring. The antlers are made of bone. As the moose age, their antlers get larger until moose reach their **prime**. After that, they start shrinking.

Learn more here!

Antlers can grow to be six feet (1.8m) wide.

Male moose use their antlers to **attract** a female moose, called a cow. Cows choose the bull with the biggest antlers. Bulls drop their antlers at the beginning of every winter.

Chapter 3

Eater of Twigs

Moose eat plants. They have to eat a lot of food because they are so big. A moose's height makes it hard to bend down to reach grass. So it eats leaves, shrubs, twigs, and bark that are higher up.

Explore links here!

16

A moose's favorite foods grow on willow, aspen, and balsam trees. Moose will eat **aquatic** plants as well. When they are searching for these plants, they can hold their breath and go underwater.

Chapter 4

Moose Mammas

Baby moose are called calves. They are usually born in May or June. Cows give birth to one calf a year. The calves can stand on their own in a day. A couple weeks after birth, calves can swim.

Complete an
activity here!

Moose mothers are
protective of their calves. A
calf drinks its mother's milk
for 6 months. It will stay with

its mother for a year. The calf is on its own before the mother gives birth to her next baby.

Making Connections

Text-to-Self

Have you ever seen a moose in the wild? If not, would you want to?

Text-to-Text

Have you read any other books about animals with antlers? How were those animals similar to moose?

Text-to-World

What do you think is the most majestic animal? Explain your answer.

Glossary

aquatic – growing or living in water.

attract – to gain the attention of.

dewlap – a loose fold of skin below the neck of certain animals.

hollow – having empty space on the inside.

hoof – the hard covering on the feet of certain animals like horses and deer.

majestic – having or showing impressive beauty, power, and/or size.

mammal – an animal that makes milk to feed its young and usually has hair or fur on its skin.

prime – the time in life when an animal is at its best. For moose bulls, this is usually between the ages of 7 and 10.

Index